636

ROBIN AND JOCELYN WILD

How Animals Work for Us

ROBIN AND JOCELYN WILD

How Animals Work for Us

FINDING-OUT BOOKS

Parents' Magazine Press
New York

Library of Congress Cataloging in Publication Data

Wild, Robin.
  How animals work for us.
  (Finding-out books)
  Published in 1973 by W. Heinemann, London, under
title: Animals at work.
  SUMMARY: Describes how man, throughout the centuries,
has trained animals to work for him in many capacities.
  1. Domestic animals–Juvenile literature.
[1. Domestic animals]   I. Wild, Jocelyn, joint author.
II. Title.
SF75.5.W54 1974   636   73-15799
ISBN 0-8193-0707-6

Contents

ONE

Animals at Work

Through the woods on a misty fall morning comes a man taking a pig for a walk. You might think he has a rather strange pet. But this is a very special little pig. Her master is a French farmer, and he has carefully trained her to find truffles. They are a kind of fungus you can eat, like mushrooms, and are very expensive to buy. Truffles grow close to the roots of oak trees, several inches under the ground, so they are hard to find. But pigs have such keen noses they can smell a truffle from twenty feet away.

Snorting and snuffling, the farmer's little pig rushes from one tree to the next. When she smells a truffle she starts busily digging. But the treasure is not for her. Her master picks it up carefully, lays it in his basket, and rewards her with a handful of corn from his pocket.

TWO

How It All Began

Hunting for truffles is just one way in which animals work for us. From earliest times all over the world men have made use of animals.

Thousands of years ago, in what we now call the Old Stone Age, much of the earth was covered with thick forest. Men looked very fierce, with shaggy hair and wild, watchful eyes. They spent most of their time roaming through the trees in a never-ending search for food. They lived on berries, roots, and the meat of wild animals, like reindeer, bison, and woolly mammoths. The hunters stalked them

with spears and stone axes, trapped them in pits, or drove whole herds over cliffs.

Men hunted animals not only for meat, but for skins to use as clothing, and for the oil to burn in their lamps. They made ornaments out of the bones, and even needles to sew with.

Men were not the only hunters prowling across the land. Hungry wolves ran in packs, using their sense of smell to track their prey. The cavemen saw how useful this could be. They would wait until the wolves had cornered an animal, and then move in to

share the spoils. This was probably the first way in which men let animals work for them.

Sometimes a man out gathering berries might come across a couple of young wolf cubs in their lair among the rocks. He would take them back to his cave and feed them. They lived with his family and played with his children, so that they grew up quite tame. The young wolves were taught to obey humans as their masters, and as soon as they were old enough the men would take them out hunting.

Imagine the hunters moving quietly through the forest. They push their way through waist-high ferns and grasses, their eyes and ears alert for the dangers hidden among the trees. The young wolves race on ahead, following the trail of a wild pig. When the hunters catch up they find the pig cornered by the snarling wolves. The men have only to close in with their spears. The hardest part of the hunt has been done for them by their wolves.

You would hardly believe that wolves had
anything to do with the meek little dogs that trot
beside old ladies in the parks today! But dogs
probably did come from wolves, after many years of
breeding. The hunting scenes painted on cave walls
show dogs running with their masters. Since then
the dog has been the animal closest to man, sharing
his home, and ready to go through all sorts of
dangers for him.

THREE

The Beginning of Farming

By the middle of the Stone Age, about ten thousand
years ago, men began domesticating wild goats and

sheep. Domesticating means catching wild animals and taming them.

Some tribes gave up the wandering life, and settled down to grow food crops for themselves and their animals. Gradually the dense forest became dotted with little clearings, carefully planted and looked after. Farming had begun.

The growing crops were a great temptation to the
animals that lived close by. Wild rabbits, pigs, and
cattle strayed in from the forest to feed on the wheat
and barley, and flocks of geese flew down.
Sometimes, instead of chasing them away, the
settlers caught them and fenced them in. After a
while they became as tame as the goats and sheep.

By this time, the Stone Age farmer had most of
the animals we have on farms today. He trained his
dogs to round up the cattle and sheep, drive them
out to pasture, and bring them back in the evening.
The dogs watched over them all the time, kept them
from straying, and protected them from wild
animals.

Other animals that joined the early farm were not
so welcome. Rats and mice robbed the precious
stores of grain. Luckily for the farmer, who knew
nothing of mousetraps or rat poisons, hungry wild
cats came to eat the rats and mice. They stayed on
the farms, and we still depend on them, all these

thousands of years later, to keep farms and mills and warehouses free of vermin.

The farmer was glad of the work done by his dogs and cats. But they could not help him with the back-breaking job of growing the crops. Digging, breaking up the clods of earth, sowing, reaping, and threshing all had to be done by hand. The farmer even had to carry everything himself, though he probably made the women of the family do most of that!

FOUR

Cattle to Pull Loads

In time the farmer saw how his big, strong cattle could make his life easier. He began to use oxen

20

to ride on and to carry heavy loads for him. He made them pull a plow to dig over his fields. The oxen did as much work in a day as the farmer had done in a week. This meant that he could farm more land.

When wheeled carts were invented it was the ox that pulled them. Soon the narrow paths widened into cart tracks. People became adventurous. They were able to travel long distances, trading their goods, and exchanging news and ideas.

The tiny settlements grew into villages and towns. With animals taking over so much of the hard work, men had time for a more varied life than that of their ancestors, the cavemen.

FIVE

Horses for Speed

In some places people learned to ride donkeys and camels, which were very hardy and needed less food and care than oxen. But it was at least a thousand years after that when men thought of catching and training the herds of wild horses that roamed free.

At first very few people dared to ride them. Instead, horses were harnessed to war chariots. They thundered down on the enemy, carrying warriors armed with bows or spears. The war chariot was such a terrifying weapon that it helped kings to conquer vast empires.

When men learned to ride on the backs of horses,
the new speed must have seemed thrilling.
Messengers on horseback took only a few days to
bring news that used to take months. New ideas
spread quicker than ever before, bringing more and
more changes to people's lives. This was the fastest
speed at which men could travel until railroads were
invented.

SIX

Horses for Power

Working horses carried packs and pulled light carts. But it was only about five hundred years ago that people found out how to harness a horse with a collar so that it could pull a heavy load with all its strength. Then horses took over much of the work that oxen had been doing for so many years. Teams of great horses pulled the farm wagons. They

worked water pumps and mills for grinding corn.
They pulled the plows that turned the soil, harrows
that raked it, and rollers that smoothed out the
lumps.

As time went on, more complicated farm
machinery was invented. Seed drills, reapers,
mowers, and combine harvesters saved the farmer's
time and helped him grow more food. They were
wonderful inventions, but quite useless without a
horse to pull them around the fields.

Horses to Carry People and Goods

Seventy-five years ago, almost everything that people needed in their daily lives was pulled along the roads by horses. Stage coaches bustled from town to town past the delivery man's cart and the lumbering wagons. Rich people traveled in their carriages, while the farmer trotted to market in his little trap.

Along the canals horses pulled the barges loaded with goods from the factories.

Down in the dark tunnels of the mines, pit ponies struggled along with heavy truckloads of coal. Some of these strong little ponies spent their whole lives underground, without ever seeing daylight or green fields.

In the cities, horse-drawn buses loaded with passengers jostled with cabs and carts of all shapes and sizes. Drivers shouted at their horses above the rattle of wheels and hoofs on the cobbled streets. Horses were as common then as cars and trucks are today.

EIGHT

Working Dogs

Through the years, dogs have been almost as useful as horses. Men have bred them for all kinds of different jobs. Little dogs were bred to catch rats and rabbits, and big dogs sometimes pulled light carts. Some kitchens used to have a dog running around inside a treadmill to turn the roasting spit or the butter churn.

In the mountains of Switzerland, enormous dogs, called St. Bernards, were bred to help rescue travelers lost in snowdrifts. Dogs are still trained for rescue

work in the mountains. When they find a man
buried under an avalanche they first dig away the
snow so that he will not suffocate, and then bark to
attract the rescue party.

On the coast of Newfoundland, fishermen used to
keep huge dogs to rescue people from drowning in
the sea.

The police find dogs a great help when they are
looking for an escaped prisoner or a child missing
from home. With its nose to the ground, the dog
can follow where someone has been just by using its
sense of smell. Policemen sometimes take their dogs

with them when they patrol the streets. A dog's nose and ears sense trouble long before the policeman notices anything wrong.

This alertness can be used to help blind people. By careful training a dog can be taught to lead its blind master around busy streets, making sure he comes to no harm.

Dogs are very good at guarding things. They will growl and bark to keep robbers away, so fierce dogs are often used to protect factories at night.

Many truck drivers have a little dog perched up in the cab with them. The dog keeps them company on long trips and guards the load when the driver stops for a meal.

NINE

Animals in War

Animals have worked for men in war as much as in peacetime, and no animal has fought for us more often than the horse. For over two thousand years, horses have carried men into battle and bravely charged the enemy. It was horses that pulled the heavy cannon and hauled all the supplies that a great army needed.

All types of horses have been used. When the knights of the Middle Ages fought, they wore such heavy armor that only very big war horses could carry the weight.

The American Indians relied on lightness and
speed, so they rode small, very fast ponies.

Soldiers on horseback were called cavalry. When the cavalry charged, it must have been a splendid and terrifying sight, with the mass of galloping horses and the brilliant uniforms of their riders.

But horses cannot stand up to tanks and machine-guns, so modern warfare has put an end to using cavalry in battle. Today horses are kept only for ceremonies and parades.

In Asia and Africa, elephants used to be taken into battle. They wore heavy suits of armor and sometimes they had swords tied to their trunks or even poisoned daggers fixed to their tusks. Soldiers rode high on their backs, hurling spears and javelins.

Dogs have worked with soldiers from early times. The Persian and Roman armies often had savage mastiffs in their front ranks. They were trained to hurl themselves on the enemy and to catch anyone who tried to run away. The dogs were so brave that they never drew back or refused to fight.

Dogs have been used to take messages during a battle, or to carry supplies on their backs for troops fighting in the mountains. Today they often work as sentries and scouts to give warning when the enemy is about.

Keen-scented dogs can be trained to go in front of advancing troops and smell out deadly mines laid in their path by the enemy. They can also be trained to go out to the battlefield and look for the wounded, particularly at night, when it would be dangerous to flash a light.

After bombing raids on towns, dogs have helped to find people buried under the rubble of bombed buildings.

Many people owe their lives to the bravery of dogs, and some dogs have even been awarded medals.

Even birds can be useful in wartime. Carrier
pigeons have flown through shell and rifle fire with
important messages fastened to their legs.
In the Second World War there was even a secret

plan to drop thousands of bats over a town by
parachute. A small time bomb was harnessed to each
bat. They were to make straight for the roofs of the
buildings, then chew through their harness and fly
away, leaving the bombs to explode.

47

TEN

Animals for Hunting

All over the world men hunt animals for food, and
sometimes just for sport, but these days most people
think hunting for sport is cruel. Men are very clever
at hunting, but it is much easier when they have the
help of a trained animal.

Sometimes a tame animal helps the hunter get close to his prey. In India, hunters often catch wild duck or partridges like this, by creeping up to them behind a trained cow. The birds are not afraid of cows, and do not suspect that this one has a man hiding behind it.

Some animals are trained to do the hunting themselves, and then let their master take what they have caught. Eastern princes hunt antelope in this way, with trained cheetahs. The cheetah races at full speed to run the antelope down, then returns obediently to its master.

Some people keep ferrets that are taught to go down burrows to catch rabbits.

Little terriers also use their noses to sniff out badgers and rabbits. They are just the right size to go down into the animal's burrow and chase it out.

Though small they are very brave and can be quite fierce.

In China and Japan, fishermen keep birds called cormorants to catch fish for them. The fisherman sends his cormorants over the edge of the boat and when they come out of the water he takes the wriggling fish from their beaks.

Of all hunting animals, dogs are some of the cleverest. There are all kinds of dogs bred to hunt in different ways. Because greyhounds are very fast they are trained to catch hares. A retriever will sit still while its master shoots birds or rabbits, and then go out and bring them back to him.

In England, red-coated huntsmen have a whole pack of hounds to follow after the scent of a fox.

ELEVEN

Animals that Entertain

Both children and grown-ups enjoy watching animals do tricks—the troupe of beautiful horses trotting around the circus ring, or the snarling lions that jump through hoops of fire. Even fleas can be taught to do tricks!

At one time performing animals were a common sight on the streets. Bears were taken from place to

place for people to watch their clumsy imitation of a dance.

You can still sometimes see a little dog acting in a Punch and Judy show, or a man with a barrel organ and a monkey that holds out a cap for your pennies.

Everyone finds it exciting to watch one animal race against another. People will risk losing a lot of money betting on which horse will win a race. Many large towns have a dog racing track where you can go at night to watch greyhounds chasing an electric hare.

Other people race pigeons. They take the birds in a basket to a place far from home, and set them free. The first pigeon to find its way home wins the race.

These days most of us think it cruel to watch

animals fighting each other. But not long ago it was a very popular amusement, and it still is in some countries.

In India, princes were sometimes entertained by a tug-of-war between two elephants. Using their tusks and trunks, the elephants struggled to pull each other over a dividing wall.

In ancient Rome great crowds filled the arenas to watch men fight against wild beasts like lions or bulls. You can still see a fight between a man and a dangerous animal at the bull rings in Spain and Mexico, where bullfighting is a national sport.

TWELVE

Animals at Work Today

Nowadays trucks and cars hurtle along our roads. The carts and carriages have gone forever. Tractors and combines rumble across fields where teams of great cart-horses once worked.

We live in an age of machines. The gasoline engine and the electric motor have taken over much of the work that animals used to do. Machines are faster and stronger and they need fewer men to

look after them. But not everyone can afford to buy
expensive machinery. So in poorer countries you can
still see animals doing the work.

There are some places where an animal does the
job even better than a machine. In the sandy deserts
of Arabia a camel can go on a long, hot journey with
its water supply stored in its body, and its splayed
feet carry it quickly and lightly over the soft sands.
The camel provides its owner with transportation,
and also with milk.

Farther north in the snows of Greenland and
Alaska, dogs are as important to the Eskimo as the
camel is to the Arab. In winter the only way an
Eskimo can travel is on his sledge, pulled by his team
of huskies. The dogs' thick coats keep them warm in
the icy blizzards, and they are so hardy that they can
keep going where many machines would break
down. Even nowadays Polar expeditions have husky
teams to haul their supplies.

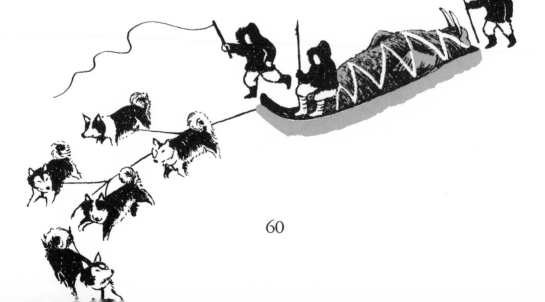

High in their snowy mountains the Tibetans could not survive without the woolly yak. They use it to ride on and to pull their carts. They drink the yak's milk and eat its meat. The Tibetans even make warm clothing and rope out of its hair. The yak is more useful to the Tibetan peasant than any tractor could be.

In the hot Indian jungle, elephants work steadily, felling trees and stacking logs. With just a few words from its rider the elephant will lean its forehead against a tree and slowly push it over. Other elephants carry huge logs on their tusks and stack them neatly in piles at the edge of the clearing. In the rainy season, elephants can go on working when bulldozers and tractors would only get stuck in the mud.

Think how much we have gained from the hard work animals have done for us so patiently through the years. They still work for us and still give us great pleasure. In return we should look after them and treat them well.

INDEX